Long Division

Practice Workbook

on Graph Paper

- With Remainders
- Without Remainders
- 1-Digit Divisors
- 2-Digit Divisors
- Decimal Quotients
- Grid Lines for Easy Number Alignment

CLAY MAZE.com

TABLE OF CONTENTS

SECTION

1

Long Division

1-Digit Divisors
2-Digit Dividends
No Remainders

```
   2 4
3 )7 2
 -6 ↓
  1 2
 -1 2
     0
```

Divide 3 into the first digit: 7 (3 can go **2** times into 7 without going over.) 2

Multiply: 3x2 = 6. Then subtract and bring down the 2.

Divide 3 into 12 (3 can go **4** times into 12 without going over.) 4

Multiply: 3x4 = 12. Then subtract.

The difference is 0 and there are no more digits to bring down, so we are done.

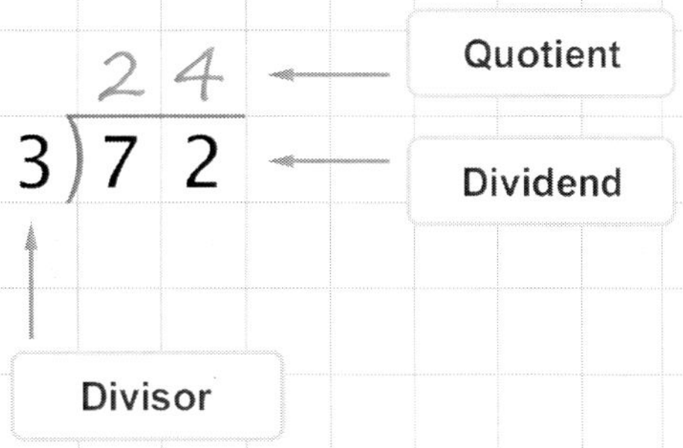

Name ______________________ Date ______________________

Divide.

1. $4\overline{)56}$

2. $2\overline{)52}$

3. $2\overline{)82}$

4. $3\overline{)48}$

5. $5\overline{)95}$

6. $4\overline{)64}$

7. $2\overline{)78}$

8. $4\overline{)68}$

9. $3\overline{)51}$

10. $4\overline{)84}$

11. $4\overline{)76}$

12. $3\overline{)84}$

13. $4\overline{)60}$

14. $3\overline{)42}$

15. $4\overline{)72}$

16. $6\overline{)84}$

Name ______________________ Date ______________

Divide.

1. $2\overline{)48}$

2. $7\overline{)91}$

3. $2\overline{)30}$

4. $2\overline{)68}$

5. $5\overline{)65}$

6. $6\overline{)96}$

7. $6\overline{)78}$

8. $2\overline{)80}$

9. $2\overline{)88}$

10. $6\overline{)90}$

11. $2\overline{)92}$

12. $3\overline{)96}$

13. $4\overline{)52}$

14. $3\overline{)63}$

15. $3\overline{)54}$

16. $4\overline{)92}$

Name ____________________ Date ____________________

Divide.

1. $5\overline{)90}$

2. $3\overline{)45}$

3. $3\overline{)81}$

4. $2\overline{)94}$

5. $3\overline{)57}$

6. $5\overline{)75}$

7. $2\overline{)36}$

8. $3\overline{)78}$

9. $3\overline{)75}$

10. $3\overline{)69}$

11. $2\overline{)84}$

12. $2\overline{)58}$

13. $2\overline{)50}$

14. $4\overline{)96}$

15. $5\overline{)70}$

16. $2\overline{)62}$

SECTION

2

Long Division
1-Digit Divisors
3-Digit Dividends
No Remainders

```
    1 3 8
5 ) 6 9 0
   -5
    1 9
   -1 5
      4 0
     -4 0
        0
```

Divide 5 into the first digit: **6** (5 can go **1** time into 6 without going over.) 1

Multiply: 5x1 = 5. Then subtract and bring down the 9.

Divide 5 into **19** (5 can go **3** times into 19 without going over.) 3

Multiply: 5x3 = 15. Then subtract and bring down the 0.

Divide 5 into **40** (5 can go **8** times into 40 without going over.) 8

Multiply: 5x8 = 40. Then subtract.

The difference is 0 and there are no more digits to bring down, so we are done.

```
      7 5
3 ) 2 2 5
   -2 1
      1 5
     -1 5
        0
```

***The first digit (2) is too small to divide by the divisor (3), so use the number formed by the first two digits:* 22.**

Divide 3 into **22** (3 can go **7** times into 22 without going over.) 7

Multiply: 3x7 = 21. Then subtract and bring down the 5.

Divide 3 into **15** (3 can go **5** times into 15 without going over.) 5

Multiply: 3x5 = 15. Then subtract.

The difference is 0 and there are no more digits to bring down, so we are done.

Name ______________________ Date ______________

Divide.

1. $3\overline{)489}$

2. $2\overline{)172}$

3. $2\overline{)724}$

4. $4\overline{)292}$

5. $7\overline{)238}$

6. $3\overline{)942}$

7. $6\overline{)552}$

8. $3\overline{)447}$

9. $9\overline{)504}$

Name ______________________ Date ______________________

Divide.

1. $4\overline{)932}$

2. $2\overline{)628}$

3. $7\overline{)588}$

4. $2\overline{)496}$

5. $6\overline{)438}$

6. $2\overline{)598}$

7. $4\overline{)776}$

8. $2\overline{)528}$

9. $8\overline{)752}$

Name ______________________ Date ______________

Divide.

1. $2\overline{)524}$

2. $9\overline{)765}$

3. $4\overline{)220}$

4. $2\overline{)918}$

5. $8\overline{)368}$

6. $6\overline{)936}$

7. $5\overline{)765}$

8. $9\overline{)882}$

9. $5\overline{)290}$

Name ______________________ Date ______________________

Divide.

1. $5\overline{)375}$

2. $2\overline{)772}$

3. $4\overline{)620}$

4. $6\overline{)492}$

5. $2\overline{)438}$

6. $2\overline{)798}$

7. $9\overline{)279}$

8. $4\overline{)716}$

9. $4\overline{)692}$

Name ______________________ Date ______________

Divide.

1. $5\overline{)820}$

2. $7\overline{)973}$

3. $2\overline{)394}$

4. $8\overline{)376}$

5. $5\overline{)705}$

6. $3\overline{)297}$

7. $2\overline{)386}$

8. $3\overline{)642}$

9. $2\overline{)984}$

Name ______________________ Date ______________________

Divide.

1. $3\overline{)624}$

2. $8\overline{)856}$

3. $2\overline{)912}$

4. $8\overline{)528}$

5. $7\overline{)224}$

6. $4\overline{)524}$

7. $6\overline{)120}$

8. $3\overline{)288}$

9. $2\overline{)530}$

Name ____________________ Date ____________________

Divide.

1. $6\overline{)150}$

2. $2\overline{)522}$

3. $6\overline{)558}$

4. $6\overline{)906}$

5. $2\overline{)584}$

6. $2\overline{)558}$

7. $4\overline{)848}$

8. $7\overline{)126}$

9. $7\overline{)511}$

Name ____________________ Date ____________________

Divide.

1. $8\overline{)152}$

2. $3\overline{)732}$

3. $9\overline{)513}$

4. $4\overline{)232}$

5. $6\overline{)780}$

6. $4\overline{)976}$

7. $4\overline{)248}$

8. $2\overline{)892}$

9. $8\overline{)672}$

Name ______________________ Date ______________

Divide.

1. $6\overline{)324}$

2. $7\overline{)231}$

3. $2\overline{)194}$

4. $9\overline{)414}$

5. $5\overline{)460}$

6. $6\overline{)342}$

7. $5\overline{)315}$

8. $3\overline{)834}$

9. $5\overline{)660}$

Name ______________________ Date ______________

Divide.

1. $5\overline{)215}$

2. $6\overline{)750}$

3. $8\overline{)144}$

4. $5\overline{)465}$

5. $9\overline{)783}$

6. $2\overline{)476}$

7. $2\overline{)304}$

8. $7\overline{)861}$

9. $3\overline{)471}$

Name ______________________________ Date ____________________

Divide.

1. $4\overline{)780}$

2. $7\overline{)476}$

3. $2\overline{)562}$

4. $3\overline{)405}$

5. $6\overline{)840}$

6. $4\overline{)756}$

7. $2\overline{)696}$

8. $2\overline{)388}$

9. $8\overline{)984}$

Name ______________________ Date ______________

Divide.

(1) $3\overline{)321}$

(2) $2\overline{)950}$

(3) $5\overline{)975}$

(4) $3\overline{)882}$

(5) $5\overline{)535}$

(6) $2\overline{)802}$

(7) $2\overline{)780}$

(8) $5\overline{)395}$

(9) $3\overline{)858}$

Name ______________________ Date ______________

Divide.

1. $6\overline{)138}$

2. $2\overline{)924}$

3. $5\overline{)685}$

4. $7\overline{)448}$

5. $4\overline{)588}$

6. $8\overline{)688}$

7. $3\overline{)441}$

8. $2\overline{)586}$

9. $4\overline{)612}$

Name ______________________ Date ______________________

Divide.

1. $2\overline{)762}$

2. $7\overline{)595}$

3. $3\overline{)822}$

4. $4\overline{)348}$

5. $7\overline{)203}$

6. $2\overline{)162}$

7. $5\overline{)640}$

8. $4\overline{)384}$

9. $5\overline{)985}$

Name ______________________ Date ______________

Divide.

1. $5\overline{)105}$

2. $4\overline{)868}$

3. $5\overline{)725}$

4. $3\overline{)324}$

5. $8\overline{)704}$

6. $2\overline{)736}$

7. $3\overline{)711}$

8. $6\overline{)102}$

9. $9\overline{)477}$

Name ______________________ Date ______________

Divide.

1. $2\overline{)208}$

2. $8\overline{)712}$

3. $7\overline{)385}$

4. $5\overline{)700}$

5. $8\overline{)120}$

6. $2\overline{)122}$

7. $4\overline{)224}$

8. $4\overline{)140}$

9. $2\overline{)346}$

Name ________________________ Date ________________

Divide.

1. $7\overline{)861}$

2. $2\overline{)746}$

3. $3\overline{)828}$

4. $4\overline{)172}$

5. $5\overline{)625}$

6. $4\overline{)984}$

7. $2\overline{)896}$

8. $6\overline{)756}$

9. $5\overline{)865}$

Name ______________________ Date ______________________

Divide.

1. $2\overline{)192}$

2. $5\overline{)495}$

3. $3\overline{)927}$

4. $3\overline{)432}$

5. $7\overline{)329}$

6. $5\overline{)440}$

7. $3\overline{)513}$

8. $9\overline{)846}$

9. $2\overline{)618}$

Name ______________________ Date ______________________

Divide.

1. $5\overline{)475}$

2. $9\overline{)612}$

3. $2\overline{)788}$

4. $2\overline{)190}$

5. $8\overline{)520}$

6. $3\overline{)978}$

7. $2\overline{)414}$

8. $5\overline{)340}$

9. $5\overline{)720}$

Name ____________________ Date ____________________

Divide.

1. $8\overline{)928}$

2. $2\overline{)506}$

3. $5\overline{)155}$

4. $4\overline{)568}$

5. $5\overline{)840}$

6. $9\overline{)486}$

7. $4\overline{)908}$

8. $3\overline{)585}$

9. $4\overline{)796}$

Name ______________________ Date ______________________

Divide.

1. $7\overline{)539}$

2. $6\overline{)876}$

3. $8\overline{)384}$

4. $3\overline{)306}$

5. $9\overline{)747}$

6. $2\overline{)574}$

7. $4\overline{)528}$

8. $4\overline{)336}$

9. $9\overline{)369}$

SECTION

3

Long Division

1-Digit Divisors
4-Digit Dividends
Remainders

```
      6 3 4 R 2
4 ) 2, 5 3 8
   -2 4
      1 3
     -1 2
        1 8
       -1 6
          2
```

***The first digit (2) is too small to divide by the divisor (4), so use the number formed by the first two digits:* 25.**

Divide 4 into **25** (4 can go **6** times into 25 without going over.) 6

Multiply: 4x6 = 24. Then subtract and bring down the 3.

Divide 4 into **13** (4 can go **3** times into 13 without going over.) 3

Multiply: 4x3 = 12. Then subtract and bring down the 8.

Divide 4 into **18** (4 can go **4** times into 18 without going over.) 4

Multiply: 4x4 = 16. Then subtract.

The difference is 2 and there are no more digits to bring down, so 2 is the remainder. R2

Name ______________________ Date ______________

Divide. Write each quotient with a remainder.

1. $8\overline{)2,382}$

2. $2\overline{)4,945}$

3. $2\overline{)6,145}$

4. $9\overline{)2,013}$

5. $6\overline{)1,436}$

6. $4\overline{)6,933}$

7. $2\overline{)1,887}$

8. $7\overline{)6,716}$

9. $3\overline{)8,527}$

Name ______________________ Date ______________________

Divide. Write each quotient with a remainder.

1. $9\overline{)2,807}$

2. $5\overline{)5,897}$

3. $4\overline{)5,534}$

4. $4\overline{)9,586}$

5. $3\overline{)2,636}$

6. $2\overline{)8,797}$

7. $3\overline{)3,272}$

8. $2\overline{)8,995}$

9. $2\overline{)7,843}$

Name ______________________ Date ______________

Divide. Write each quotient with a remainder.

1. $2\overline{)7,033}$

2. $3\overline{)1,607}$

3. $2\overline{)2,641}$

4. $4\overline{)8,831}$

5. $7\overline{)5,496}$

6. $8\overline{)1,455}$

7. $9\overline{)4,814}$

8. $3\overline{)3,029}$

9. $7\overline{)9,728}$

Name ________________________ Date ________________

Divide. Write each quotient with a remainder.

1. $2\overline{)2,793}$

2. $8\overline{)4,201}$

3. $9\overline{)2,966}$

4. $2\overline{)9,731}$

5. $7\overline{)4,108}$

6. $8\overline{)9,497}$

7. $4\overline{)2,807}$

8. $2\overline{)6,951}$

9. $8\overline{)8,155}$

Name ______________________ Date ______________

Divide. Write each quotient with a remainder.

1. $4\overline{)2,699}$

2. $5\overline{)1,266}$

3. $7\overline{)7,529}$

4. $2\overline{)5,811}$

5. $7\overline{)1,171}$

6. $2\overline{)2,981}$

7. $3\overline{)5,752}$

8. $2\overline{)1,845}$

9. $3\overline{)1,750}$

Name ______________________ Date ______________

Divide. Write each quotient with a remainder.

1. $9\overline{)1{,}049}$

2. $2\overline{)1{,}757}$

3. $4\overline{)7{,}599}$

4. $3\overline{)3{,}656}$

5. $6\overline{)9{,}389}$

6. $3\overline{)6{,}503}$

7. $2\overline{)9{,}513}$

8. $2\overline{)6{,}597}$

9. $3\overline{)6{,}055}$

Name ______________________________ Date ____________________

Divide. Write each quotient with a remainder.

1. $6\overline{)2,327}$

2. $4\overline{)7,277}$

3. $2\overline{)9,597}$

4. $5\overline{)3,943}$

5. $2\overline{)3,959}$

6. $7\overline{)5,941}$

7. $2\overline{)1,345}$

8. $3\overline{)9,367}$

9. $3\overline{)1,859}$

Name ______________________ Date ______________

Divide. Write each quotient with a remainder.

1. $5\overline{)9,239}$

2. $3\overline{)1,228}$

3. $5\overline{)3,097}$

4. $3\overline{)4,736}$

5. $2\overline{)3,631}$

6. $4\overline{)4,171}$

7. $5\overline{)2,351}$

8. $9\overline{)1,870}$

9. $2\overline{)8,049}$

Name ______________________ Date ______________________

Divide. Write each quotient with a remainder.

1. $7\overline{)7,960}$

2. $2\overline{)1,081}$

3. $4\overline{)1,993}$

4. $4\overline{)6,787}$

5. $8\overline{)4,506}$

6. $6\overline{)5,299}$

7. $7\overline{)5,700}$

8. $4\overline{)5,943}$

9. $2\overline{)8,531}$

Name ______________________ Date ______________________

Divide. Write each quotient with a remainder.

1. $3\overline{)4,108}$

2. $7\overline{)4,448}$

3. $2\overline{)8,223}$

4. $2\overline{)7,299}$

5. $8\overline{)7,831}$

6. $4\overline{)7,789}$

7. $6\overline{)4,113}$

8. $5\overline{)3,081}$

9. $5\overline{)4,272}$

Name ______________________________ Date ____________________

Divide. Write each quotient with a remainder.

1. $4\overline{)6,971}$

2. $5\overline{)8,118}$

3. $4\overline{)2,793}$

4. $8\overline{)9,545}$

5. $3\overline{)3,355}$

6. $6\overline{)2,641}$

7. $7\overline{)1,608}$

8. $2\overline{)7,933}$

9. $2\overline{)2,127}$

Name ______________________ Date ______________________

Divide. Write each quotient with a remainder.

1. $4\overline{)1,866}$

2. $2\overline{)2,089}$

3. $3\overline{)5,042}$

4. $3\overline{)5,563}$

5. $6\overline{)8,645}$

6. $4\overline{)9,683}$

7. $5\overline{)7,268}$

8. $2\overline{)6,201}$

9. $7\overline{)8,641}$

Name ________________________________ Date ____________________

Divide. Write each quotient with a remainder.

1. $3\overline{)7{,}058}$

2. $2\overline{)9{,}015}$

3. $9\overline{)8{,}759}$

4. $3\overline{)1{,}970}$

5. $2\overline{)2{,}613}$

6. $5\overline{)7{,}207}$

7. $3\overline{)3{,}896}$

8. $4\overline{)7{,}515}$

9. $3\overline{)9{,}070}$

Name ______________________ Date ______________________

Divide. Write each quotient with a remainder.

1. $7\overline{)6,236}$

2. $2\overline{)4,775}$

3. $3\overline{)3,091}$

4. $7\overline{)4,838}$

5. $3\overline{)4,315}$

6. $5\overline{)7,182}$

7. $2\overline{)2,497}$

8. $9\overline{)4,351}$

9. $2\overline{)8,951}$

Name ______________________ Date ______________

Divide. Write each quotient with a remainder.

1. $4\overline{)9{,}819}$

2. $3\overline{)4{,}376}$

3. $5\overline{)3{,}148}$

4. $2\overline{)8{,}745}$

5. $5\overline{)9{,}544}$

6. $9\overline{)7{,}868}$

7. $5\overline{)9{,}392}$

8. $2\overline{)2{,}987}$

9. $7\overline{)3{,}613}$

Name ____________________ Date ____________________

Divide. Write each quotient with a remainder.

1. $3\overline{)7,684}$

2. $9\overline{)5,498}$

3. $2\overline{)3,479}$

4. $5\overline{)3,958}$

5. $2\overline{)7,475}$

6. $6\overline{)4,257}$

7. $4\overline{)6,162}$

8. $3\overline{)8,396}$

9. $8\overline{)2,122}$

Name ______________________ Date ______________

Divide. Write each quotient with a remainder.

1. $2\overline{)2{,}171}$

2. $4\overline{)7{,}209}$

3. $3\overline{)8{,}801}$

4. $2\overline{)5{,}397}$

5. $3\overline{)8{,}683}$

6. $2\overline{)1{,}441}$

7. $4\overline{)4{,}254}$

8. $7\overline{)5{,}227}$

9. $3\overline{)4{,}979}$

Name ______________________ Date ______________________

Divide. Write each quotient with a remainder.

1. $4\overline{)1,631}$

2. $6\overline{)2,915}$

3. $5\overline{)9,011}$

4. $4\overline{)1,033}$

5. $8\overline{)2,212}$

6. $4\overline{)3,485}$

7. $9\overline{)8,463}$

8. $3\overline{)4,630}$

9. $9\overline{)2,728}$

Name ____________________ Date ____________________

Divide. Write each quotient with a remainder.

1. $3\overline{)8,663}$

2. $6\overline{)5,321}$

3. $9\overline{)3,617}$

4. $3\overline{)4,534}$

5. $2\overline{)7,845}$

6. $3\overline{)7,579}$

7. $4\overline{)8,877}$

8. $9\overline{)7,512}$

9. $3\overline{)3,542}$

Name ______________________ Date ______________

Divide. Write each quotient with a remainder.

1. $2\overline{)9,103}$

2. $9\overline{)1,100}$

3. $8\overline{)5,587}$

4. $5\overline{)6,839}$

5. $8\overline{)2,263}$

6. $6\overline{)1,451}$

7. $7\overline{)7,030}$

8. $2\overline{)6,339}$

9. $2\overline{)3,411}$

Name ______________________ Date ______________

Divide. Write each quotient with a remainder.

1. $4\overline{)9,141}$

2. $9\overline{)9,634}$

3. $8\overline{)2,339}$

4. $8\overline{)9,177}$

5. $2\overline{)4,519}$

6. $8\overline{)4,970}$

7. $4\overline{)8,261}$

8. $2\overline{)5,011}$

9. $5\overline{)9,749}$

SECTION

4 Long Division

2-Digit Divisors
4-Digit Dividends
No Remainders

```
        3 3 8
1 3)4, 3 9 4
   -3 9 ↓  |
     4 9   |
    -3 9   ↓
     1 0 4
    -1 0 4
         0
```

***The first digit (4) is too small to divide by the divisor (13), so use the number formed by the first two digits:* 43.**

Divide 13 into **43** (13 can go **3** times into 43 without going over.) 3

Multiply: 13x3 = 39. Then subtract and bring down the 9.

Divide 13 into **49** (13 can go **3** times into 49 without going over.) 3

Multiply: 13x3 = 39. Then subtract and bring down the 4.

Divide 13 into **104** (13 can go **8** times into 104 without going over.) 8

Multiply: 13x8 = 104. Then subtract.

The difference is 0 and there are no more digits to bring down, so we are done.

Name ______________________________ Date ____________________

Divide.

1. $21\overline{)6{,}531}$

2. $92\overline{)9{,}476}$

3. $20\overline{)7{,}340}$

4. $45\overline{)5{,}490}$

5. $13\overline{)4{,}394}$

6. $83\overline{)9{,}296}$

7. $64\overline{)8{,}128}$

8. $93\overline{)9{,}765}$

9. $39\overline{)7{,}878}$

Name ______________________ Date ______________________

Divide.

1. $90\overline{)9{,}810}$

2. $19\overline{)8{,}873}$

3. $38\overline{)4{,}028}$

4. $75\overline{)8{,}550}$

5. $77\overline{)8{,}470}$

6. $21\overline{)7{,}119}$

7. $30\overline{)8{,}280}$

8. $17\overline{)7{,}752}$

9. $42\overline{)5{,}544}$

Name ____________________ Date ____________________

Divide.

1. $53\overline{)7,314}$

2. $26\overline{)8,762}$

3. $51\overline{)5,457}$

4. $14\overline{)6,062}$

5. $92\overline{)9,384}$

6. $37\overline{)6,105}$

7. $45\overline{)8,775}$

8. $23\overline{)8,027}$

9. $74\overline{)9,250}$

Name ______________________ Date ______________

Divide.

1 $80\overline{)8,720}$

2 $43\overline{)9,890}$

3 $22\overline{)9,504}$

4 $87\overline{)9,048}$

5 $66\overline{)9,570}$

6 $43\overline{)4,816}$

7 $36\overline{)8,028}$

8 $25\overline{)8,275}$

9 $42\overline{)7,182}$

Name ______________________________ Date ______________

Divide.

1. $51\overline{)8,823}$

2. $43\overline{)9,761}$

3. $35\overline{)4,270}$

4. $22\overline{)8,316}$

5. $42\overline{)9,114}$

6. $38\overline{)8,322}$

7. $76\overline{)9,424}$

8. $22\overline{)6,908}$

9. $25\overline{)2,875}$

Name ________________________ Date ________________

Divide.

1. $34\overline{)8{,}806}$

2. $81\overline{)8{,}667}$

3. $28\overline{)6{,}048}$

4. $56\overline{)7{,}392}$

5. $73\overline{)9{,}417}$

6. $21\overline{)6{,}552}$

7. $28\overline{)5{,}824}$

8. $53\overline{)6{,}042}$

9. $32\overline{)7{,}680}$

Name ________________ Date ________________

Divide.

1. $14\overline{)9,114}$

2. $21\overline{)8,904}$

3. $81\overline{)8,505}$

4. $37\overline{)6,771}$

5. $71\overline{)9,159}$

6. $13\overline{)4,953}$

7. $76\overline{)9,348}$

8. $31\overline{)7,595}$

9. $59\overline{)8,142}$

Name ______________________ Date ______________

Divide.

1. $78\overline{)9,516}$

2. $24\overline{)5,568}$

3. $21\overline{)3,885}$

4. $31\overline{)5,270}$

5. $55\overline{)8,965}$

6. $24\overline{)6,072}$

7. $13\overline{)4,641}$

8. $26\overline{)5,434}$

9. $47\overline{)6,627}$

Name ______________________ Date ______________

Divide.

1. $79\overline{)8,848}$

2. $17\overline{)9,163}$

3. $62\overline{)6,572}$

4. $84\overline{)9,912}$

5. $75\overline{)9,150}$

6. $27\overline{)8,127}$

7. $60\overline{)6,300}$

8. $25\overline{)4,575}$

9. $46\overline{)9,430}$

Name ______________________ Date ______________________

Divide.

1. $23\overline{)7,429}$

2. $15\overline{)2,745}$

3. $61\overline{)9,516}$

4. $33\overline{)7,062}$

5. $28\overline{)4,340}$

6. $30\overline{)9,180}$

7. $22\overline{)4,994}$

8. $47\overline{)9,964}$

9. $91\overline{)9,464}$

Name ______________________ Date ______________

Divide.

1. $66\overline{)7,854}$

2. $24\overline{)4,104}$

3. $36\overline{)8,136}$

4. $78\overline{)9,672}$

5. $16\overline{)3,648}$

6. $13\overline{)2,379}$

7. $30\overline{)7,680}$

8. $35\overline{)6,055}$

9. $29\overline{)8,062}$

Name ______________________ Date ______________

Divide.

1. $30\overline{)6,570}$

2. $56\overline{)6,272}$

3. $16\overline{)5,904}$

4. $88\overline{)9,152}$

5. $68\overline{)9,588}$

6. $58\overline{)7,598}$

7. $15\overline{)5,325}$

8. $85\overline{)8,840}$

9. $27\overline{)6,750}$

Name ______________________ Date ______________________

Divide.

1. $55\overline{)6,490}$

2. $48\overline{)8,544}$

3. $42\overline{)9,702}$

4. $28\overline{)6,692}$

5. $24\overline{)6,048}$

6. $32\overline{)6,496}$

7. $40\overline{)9,880}$

8. $61\overline{)8,235}$

9. $39\overline{)4,524}$

Name ______________________ Date ______________________

Divide.

1. $31\overline{)7,099}$

2. $74\overline{)7,770}$

3. $34\overline{)4,794}$

4. $14\overline{)7,210}$

5. $57\overline{)7,524}$

6. $23\overline{)5,175}$

7. $48\overline{)5,232}$

8. $29\overline{)5,713}$

9. $29\overline{)7,424}$

Name ____________________ Date ____________________

Divide.

1. $40\overline{)9,320}$

2. $17\overline{)2,822}$

3. $82\overline{)9,676}$

4. $92\overline{)9,752}$

5. $29\overline{)7,308}$

6. $31\overline{)4,867}$

7. $39\overline{)8,424}$

8. $14\overline{)6,790}$

9. $47\overline{)7,567}$

Name ______________________ Date ______________

Divide.

1. $28\overline{)8,120}$

2. $58\overline{)8,410}$

3. $71\overline{)8,449}$

4. $73\overline{)9,928}$

5. $24\overline{)5,832}$

6. $17\overline{)9,163}$

7. $15\overline{)2,745}$

8. $62\overline{)7,378}$

9. $35\overline{)9,415}$

Name ______________________ Date ______________

Divide.

1. $51\overline{)6,375}$

2. $35\overline{)8,120}$

3. $58\overline{)7,076}$

4. $36\overline{)8,316}$

5. $73\overline{)8,468}$

6. $25\overline{)8,150}$

7. $73\overline{)9,344}$

8. $47\overline{)9,541}$

9. $26\overline{)3,926}$

Name ______________________ Date ______________________

Divide.

1. $25\overline{)7,925}$

2. $13\overline{)7,462}$

3. $28\overline{)8,764}$

4. $57\overline{)8,322}$

5. $73\overline{)7,738}$

6. $50\overline{)9,150}$

7. $19\overline{)8,835}$

8. $41\overline{)6,314}$

9. $19\overline{)8,778}$

Name ______________________ Date ______________

Divide.

1) $52\overline{)5,564}$

2) $30\overline{)8,520}$

3) $88\overline{)9,064}$

4) $76\overline{)8,740}$

5) $23\overline{)7,429}$

6) $48\overline{)8,784}$

7) $22\overline{)4,994}$

8) $54\overline{)7,830}$

9) $29\overline{)7,830}$

Name ______________________ Date ______________________

Divide.

1. $24\overline{)8,256}$

2. $35\overline{)8,400}$

3. $14\overline{)1,848}$

4. $91\overline{)9,464}$

5. $15\overline{)2,745}$

6. $18\overline{)5,274}$

7. $52\overline{)6,864}$

8. $27\overline{)3,618}$

9. $23\overline{)7,958}$

Name ______________________ Date ______________

Divide.

1. $45\overline{)7,785}$

2. $33\overline{)9,240}$

3. $13\overline{)4,277}$

4. $17\overline{)8,415}$

5. $29\overline{)7,685}$

6. $61\overline{)9,638}$

7. $76\overline{)9,728}$

8. $20\overline{)9,840}$

9. $43\overline{)8,342}$

SECTION

5

Long Division

1-Digit Divisors
3-Digit Dividends
Decimal Quotients

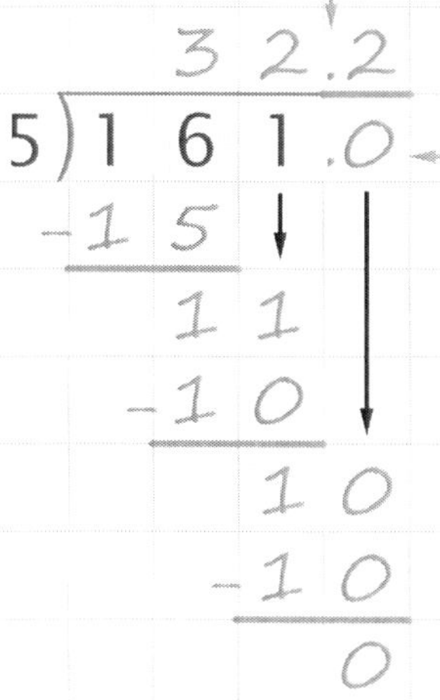

***The first digit (1) is too small to divide by the divisor (5), so use the number formed by the first two digits:* 16.**

Divide 5 into **16** (5 can go **3** times into 16 without going over.) 3

Multiply: 5x3 = 15. Then subtract and bring down the 1.

Divide 5 into **11** (5 can go **2** times into 11 without going over.) 2

Multiply: 5x2 = 10. Then subtract.

The difference is 1 and there are no more digits to bring down.
Get the decimal part of the quotient:
Place a **.0** at the end of the dividend.
Place a **.** right above it in the quotient. (Line up the decimals.)
Bring down that 0 and continue.

Divide 5 into **10** (5 can go **2** times into 10 without going over.) .2

Multiply: 5x2 = 10. Then subtract.

The difference is 0 and there are no more digits to bring down, so we are done.

Name ______________________ Date ______________

Divide. Write each quotient using a decimal instead of a remainder.

1. $5\overline{)912}$

2. $5\overline{)543}$

3. $2\overline{)261}$

4. $5\overline{)836}$

5. $5\overline{)743}$

6. $5\overline{)534}$

7. $2\overline{)655}$

8. $2\overline{)713}$

9. $5\overline{)222}$

Name ______________________ Date ______________________

Divide. Write each quotient using a decimal instead of a remainder.

1. $2\overline{)215}$

2. $2\overline{)565}$

3. $4\overline{)262}$

4. $5\overline{)714}$

5. $5\overline{)274}$

6. $5\overline{)908}$

7. $5\overline{)376}$

8. $5\overline{)298}$

9. $5\overline{)953}$

Name ______________________ Date ______________

Divide. Write each quotient using a decimal instead of a remainder.

1. $5\overline{)501}$

2. $5\overline{)604}$

3. $4\overline{)926}$

4. $4\overline{)994}$

5. $2\overline{)491}$

6. $5\overline{)654}$

7. $6\overline{)393}$

8. $8\overline{)932}$

9. $2\overline{)829}$

Name ______________________ Date ______________________

Divide. Write each quotient using a decimal instead of a remainder.

1. $5\overline{)602}$

2. $2\overline{)467}$

3. $5\overline{)347}$

4. $8\overline{)956}$

5. $2\overline{)767}$

6. $5\overline{)678}$

7. $5\overline{)182}$

8. $6\overline{)513}$

9. $4\overline{)250}$

Name ______________________ Date ______________

Divide. Write each quotient using a decimal instead of a remainder.

1. $4\overline{)474}$

2. $8\overline{)452}$

3. $4\overline{)614}$

4. $5\overline{)281}$

5. $2\overline{)429}$

6. $5\overline{)427}$

7. $2\overline{)287}$

8. $8\overline{)972}$

9. $5\overline{)831}$

Name ______________________ Date ______________________

Divide. Write each quotient using a decimal instead of a remainder.

1. $2\overline{)479}$

2. $2\overline{)777}$

3. $2\overline{)847}$

4. $5\overline{)438}$

5. $8\overline{)580}$

6. $5\overline{)286}$

7. $6\overline{)369}$

8. $6\overline{)231}$

9. $5\overline{)549}$

Name ______________________ Date ______________

Divide. Write each quotient using a decimal instead of a remainder.

1. $5\overline{)722}$

2. $2\overline{)375}$

3. $2\overline{)663}$

4. $5\overline{)507}$

5. $2\overline{)551}$

6. $5\overline{)386}$

7. $2\overline{)331}$

8. $8\overline{)516}$

9. $4\overline{)846}$

Name ______________________ Date ______________

Divide. Write each quotient using a decimal instead of a remainder.

1. $5\overline{)184}$

2. $2\overline{)545}$

3. $5\overline{)864}$

4. $2\overline{)169}$

5. $4\overline{)962}$

6. $6\overline{)495}$

7. $5\overline{)932}$

8. $5\overline{)156}$

9. $2\overline{)795}$

Name ______________________ Date ______________

Divide. Write each quotient using a decimal instead of a remainder.

1. $2\overline{)273}$

2. $5\overline{)257}$

3. $6\overline{)849}$

4. $5\overline{)489}$

5. $5\overline{)311}$

6. $2\overline{)903}$

7. $5\overline{)919}$

8. $2\overline{)801}$

9. $2\overline{)757}$

Name ______________________ Date ______________

Divide. Write each quotient using a decimal instead of a remainder.

1. $4\overline{)506}$

2. $5\overline{)331}$

3. $8\overline{)332}$

4. $5\overline{)497}$

5. $6\overline{)213}$

6. $2\overline{)343}$

7. $5\overline{)334}$

8. $8\overline{)364}$

9. $5\overline{)974}$

Name ____________________ Date ____________________

Divide. Write each quotient using a decimal instead of a remainder.

1. $5\overline{)812}$

2. $4\overline{)654}$

3. $5\overline{)437}$

4. $5\overline{)583}$

5. $2\overline{)885}$

6. $8\overline{)684}$

7. $4\overline{)574}$

8. $2\overline{)913}$

9. $2\overline{)687}$

Name ______________________ Date ______________

Divide. Write each quotient using a decimal instead of a remainder.

1. $5\overline{)209}$

2. $5\overline{)571}$

3. $4\overline{)102}$

4. $5\overline{)192}$

5. $2\overline{)105}$

6. $2\overline{)541}$

7. $5\overline{)207}$

8. $5\overline{)856}$

9. $6\overline{)603}$

Name ______________________ Date ______________

Divide. Write each quotient using a decimal instead of a remainder.

1. $5\overline{)526}$

2. $2\overline{)353}$

3. $5\overline{)676}$

4. $4\overline{)878}$

5. $5\overline{)354}$

6. $4\overline{)314}$

7. $2\overline{)381}$

8. $4\overline{)986}$

9. $2\overline{)639}$

Name ______________________ Date ______________________

Divide. Write each quotient using a decimal instead of a remainder.

1. $5\overline{)198}$

2. $2\overline{)325}$

3. $5\overline{)186}$

4. $5\overline{)472}$

5. $8\overline{)532}$

6. $2\overline{)869}$

7. $4\overline{)234}$

8. $2\overline{)101}$

9. $2\overline{)571}$

Name ______________________________ Date ____________________

Divide. Write each quotient using a decimal instead of a remainder.

1. $5\overline{)536}$

2. $4\overline{)910}$

3. $5\overline{)918}$

4. $5\overline{)442}$

5. $4\overline{)170}$

6. $8\overline{)500}$

7. $2\overline{)113}$

8. $8\overline{)796}$

9. $5\overline{)426}$

Name ______________________________ Date ____________________

Divide. Write each quotient using a decimal instead of a remainder.

1. $8\overline{)828}$

2. $2\overline{)483}$

3. $5\overline{)109}$

4. $8\overline{)996}$

5. $5\overline{)609}$

6. $4\overline{)742}$

7. $8\overline{)612}$

8. $5\overline{)518}$

9. $6\overline{)711}$

Name ______________________ Date ______________

Divide. Write each quotient using a decimal instead of a remainder.

1. $6\overline{)939}$

2. $2\overline{)495}$

3. $2\overline{)111}$

4. $4\overline{)810}$

5. $5\overline{)786}$

6. $8\overline{)380}$

7. $5\overline{)569}$

8. $4\overline{)282}$

9. $2\overline{)259}$

Name ____________________ Date ____________________

Divide. Write each quotient using a decimal instead of a remainder.

1. $4\overline{)566}$

2. $2\overline{)769}$

3. $5\overline{)563}$

4. $6\overline{)105}$

5. $5\overline{)539}$

6. $5\overline{)408}$

7. $4\overline{)502}$

8. $2\overline{)861}$

9. $2\overline{)781}$

Name ______________________ Date ______________

Divide. Write each quotient using a decimal instead of a remainder.

1) $5\overline{)188}$

2) $2\overline{)427}$

3) $2\overline{)735}$

4) $6\overline{)345}$

5) $5\overline{)538}$

6) $8\overline{)868}$

7) $2\overline{)841}$

8) $2\overline{)515}$

9) $5\overline{)422}$

Name ______________________ Date ______________

Divide. Write each quotient using a decimal instead of a remainder.

1. $5\overline{)564}$

2. $2\overline{)739}$

3. $5\overline{)688}$

4. $6\overline{)453}$

5. $5\overline{)791}$

6. $2\overline{)233}$

7. $8\overline{)476}$

8. $4\overline{)902}$

9. $5\overline{)869}$

Name ______________________ Date ______________

Divide. Write each quotient using a decimal instead of a remainder.

1. $5\overline{)622}$

2. $2\overline{)605}$

3. $2\overline{)347}$

4. $4\overline{)778}$

5. $5\overline{)251}$

6. $5\overline{)597}$

7. $5\overline{)677}$

8. $4\overline{)182}$

9. $6\overline{)543}$

SECTION

A

Answer Key

Answers to problems.

Answer Key

PAGE 2

14	26	41	16
19	16	39	17
17	21	19	28
15	14	18	14

PAGE 3

24	13	15	34
13	16	13	40
44	15	46	32
13	21	18	23

PAGE 4

18	15	27	47
19	15	18	26
25	23	42	29
25	24	14	31

PAGE 6

163	86	362
73	34	314
92	149	56

PAGE 7

233	314	84
248	73	299
194	264	94

PAGE 8

262	85	55
459	46	156
153	98	58

PAGE 9

75	386	155
82	219	399
31	179	173

PAGE 10

164	139	197
47	141	99
193	214	492

PAGE 11

208	107	456
66	32	131
20	96	265

PAGE 12

25	261	93
151	292	279
212	18	73

Answer Key

PAGE 13

19	244	57
58	130	244
62	446	84

PAGE 14

54	33	97
46	92	57
63	278	132

PAGE 15

43	125	18
93	87	238
152	123	157

PAGE 16

195	68	281
135	140	189
348	194	123

PAGE 17

107	475	195
294	107	401
390	79	286

PAGE 18

23	462	137
64	147	86
147	293	153

PAGE 19

381	85	274
87	29	81
128	96	197

PAGE 20

21	217	145
108	88	368
237	17	53

PAGE 21

104	89	55
140	15	61
56	35	173

PAGE 22

123	373	276
43	125	246
448	126	173

PAGE 23

96	99	309
144	47	88
171	94	309

PAGE 24

95	68	394
95	65	326
207	68	144

PAGE 25

116	253	31
142	168	54
227	195	199

PAGE 26

77	146	48
102	83	287
132	84	41

PAGE 28

297 R6	2472 R1	3072 R1
223 R6	239 R2	1733 R1
943 R1	959 R3	2842 R1

PAGE 29

311 R8	1179 R2	1383 R2
2396 R2	878 R2	4398 R1
1090 R2	4497 R1	3921 R1

PAGE 30

3516 R1	535 R2	1320 R1
2207 R3	785 R1	181 R7
534 R8	1009 R2	1389 R5

PAGE 31

1396 R1	525 R1	329 R5
4865 R1	586 R6	1187 R1
701 R3	3475 R1	1019 R3

PAGE 32

674 R3	253 R1	1075 R4
2905 R1	167 R2	1490 R1
1917 R1	922 R1	583 R1

PAGE 33

116 R5	878 R1	1899 R3
1218 R2	1564 R5	2167 R2
4756 R1	3298 R1	2018 R1

Answer Key

PAGE 34

387 R5	1819 R1	4798 R1
788 R3	1979 R1	848 R5
672 R1	3122 R1	619 R2

PAGE 35

1847 R4	409 R1	619 R2
1578 R2	1815 R1	1042 R3
470 R1	207 R7	4024 R1

PAGE 36

1137 R1	540 R1	498 R1
1696 R3	563 R2	883 R1
814 R2	1485 R3	4265 R1

PAGE 37

1369 R1	635 R3	4111 R1
3649 R1	978 R7	1947 R1
685 R3	616 R1	854 R2

PAGE 38

1742 R3	1623 R3	698 R1
1193 R1	1118 R1	440 R1
229 R5	3966 R1	1063 R1

PAGE 39

466 R2	1044 R1	1680 R2
1854 R1	1440 R5	2420 R3
1453 R3	3100 R1	1234 R3

PAGE 40

2352 R2	4507 R1	973 R2
656 R2	1306 R1	1441 R2
1298 R2	1878 R3	3023 R1

PAGE 41

890 R6	2387 R1	1030 R1
691 R1	1438 R1	1436 R2
1248 R1	483 R4	4475 R1

PAGE 42

2454 R3	1458 R2	629 R3
4372 R1	1908 R4	874 R2
1878 R2	1493 R1	516 R1

PAGE 43

2561 R1	610 R8	1739 R1
791 R3	3737 R1	709 R3
1540 R2	2798 R2	265 R2

PAGE 44

1085 R1	1802 R1	2933 R2
2698 R1	2894 R1	720 R1
1063 R2	746 R5	1659 R2

PAGE 45

407 R3	485 R5	1802 R1
258 R1	276 R4	871 R1
940 R3	1543 R1	303 R1

PAGE 46

2887 R2	886 R5	401 R8
1511 R1	3922 R1	2526 R1
2219 R1	834 R6	1180 R2

PAGE 47

4551 R1	122 R2	698 R3
1367 R4	282 R7	241 R5
1004 R2	3169 R1	1705 R1

PAGE 48

2285 R1	1070 R4	292 R3
1147 R1	2259 R1	621 R2
2065 R1	2505 R1	1949 R4

PAGE 50

311	103	367
122	338	112
127	105	202

PAGE 51

109	467	106
114	110	339
276	456	132

PAGE 52

138	337	107
433	102	165
195	349	125

PAGE 53

109	230	432
104	145	112
223	331	171

PAGE 54

173	227	122
378	217	219
124	314	115

Answer Key

PAGE 55

259	107	216
132	129	312
208	114	240

PAGE 56

651	424	105
183	129	381
123	245	138

PAGE 57

122	232	185
170	163	253
357	209	141

PAGE 58

112	539	106
118	122	301
105	183	205

PAGE 59

323	183	156
214	155	306
227	212	104

PAGE 60

119	171	226
124	228	183
256	173	278

PAGE 61

219	112	369
104	141	131
355	104	250

PAGE 62

118	178	231
239	252	203
247	135	116

PAGE 63

229	105	141
515	132	225
109	197	256

PAGE 64

233	166	118
106	252	157
216	485	161

Answer Key

PAGE 65

290	145	119
136	243	539
183	119	269

PAGE 66

125	232	122
231	116	326
128	203	151

PAGE 67

317	574	313
146	106	183
465	154	462

PAGE 68

107	284	103
115	323	183
227	145	270

PAGE 69

344	240	132
104	183	293
132	134	346

PAGE 70

173	280	329
495	265	158
128	492	194

PAGE 72

182.4	108.6	130.5
167.2	148.6	106.8
327.5	356.5	44.4

PAGE 73

107.5	282.5	65.5
142.8	54.8	181.6
75.2	59.6	190.6

PAGE 74

100.2	120.8	231.5
248.5	245.5	130.8
65.5	116.5	414.5

PAGE 75

120.4	233.5	69.4
119.5	383.5	135.6
36.4	85.5	62.5

PAGE 76

118.5	56.5	153.5
56.2	214.5	85.4
143.5	121.5	166.2

PAGE 77

239.5	388.5	423.5
87.6	72.5	57.2
61.5	38.5	109.8

PAGE 78

144.4	187.5	331.5
101.4	275.5	77.2
165.5	64.5	211.5

PAGE 79

36.8	272.5	172.8
84.5	240.5	82.5
186.4	31.2	397.5

PAGE 80

136.5	51.4	141.5
97.8	62.2	451.5
183.8	400.5	378.5

PAGE 81

126.5	66.2	41.5
99.4	35.5	171.5
66.8	45.5	194.8

PAGE 82

162.4	163.5	87.4
116.6	442.5	85.5
143.5	456.5	343.5

PAGE 83

41.8	114.2	25.5
38.4	52.5	270.5
41.4	171.2	100.5

PAGE 84

105.2	176.5	135.2
219.5	70.8	78.5
190.5	246.5	319.5

PAGE 85

39.6	162.5	37.2
94.4	66.5	434.5
58.5	50.5	285.5

PAGE 86

107.2	227.5	183.6
88.4	42.5	62.5
56.5	99.5	85.2

PAGE 87

103.5	241.5	21.8
124.5	121.8	185.5
76.5	103.6	118.5

PAGE 88

156.5	247.5	55.5
202.5	157.2	47.5
113.8	70.5	129.5

PAGE 89

141.5	384.5	112.6
17.5	107.8	81.6
125.5	430.5	390.5

PAGE 90

37.6	213.5	367.5
57.5	107.6	108.5
420.5	257.5	84.4

PAGE 91

112.8	369.5	137.6
75.5	158.2	116.5
59.5	225.5	173.8

PAGE 92

124.4	302.5	173.5
194.5	50.2	119.4
135.4	45.5	90.5

We hope you found our book helpful.
Please consider leaving us a review.

Made in United States
North Haven, CT
05 November 2026

21772479R00059